Michael Phelps

By Jeffrey Zuehlke

LERNERSPORTS / Minneapolis

This book is available in two editions:
Library binding by LernerSports
Soft cover by First Avenue Editions
Imprints of Lerner Publishing Group
241 First Avenue North
Minneapolis, MN 55401 U.S.A.

Website address: www.lernerbooks.com

Library of Congress Cataloging-in-Publication Data

Zuehlke, Jeffrey, 1968–
Michael Phelps / by Jeffrey Zuehlke.
p. cm. — (Amazing athletes)
Includes index.
ISBN: 0-8225-2431-7 (lib. bdg. : alk. paper)
ISBN: 0-8225-2631-X (pbk. : alk. paper)
1. Phelps, Michael, 1985—Juvenile literature. 2. Swimmers—United States—Biography—Juvenile literature. 3. Olympic Games (28th : 2004 : Athens, Greece)—Juvenile literature. I. Title.
GV838.P54Z84 2005
797.2'1'092—dc22 2004021742

Manufactured in the United States of America
1 2 3 4 5 6 – DP – 10 09 08 07 06 05

Table of Contents

Michael dives off the starting block in the 400-meter **individual medley** race at the 2004 Olympics Games.

Good as Gold

Michael Phelps stood on the **starting block.** He stretched his long arms down in front of him.

"Take your marks!" said the voice on the loudspeaker.

"Honk!" the starting buzzer sounded.

Michael leaped into the water. He sped beneath the surface like a submarine. Finally, he came up and began swimming the hardest **stroke,** the **butterfly.**

Up and down, up and down, in and out of the water. He was in the lead!

Michael was competing against seven other swimmers. He was racing in the **final** of the 400-meter individual medley at the 2004 Olympic Games. In this race, swimmers use four different strokes, one after the other.

Olympic-sized pools are 50 meters long. So in a 100-meter race, swimmers have to cross the pool twice. And in a 200-meter race, swimmers have to cross the pool four times.

Michael was just nineteen years old. But he was the heavy **favorite** to win the **event.**

After 100 meters, Michael switched to the **backstroke.** He was ahead of the second-place swimmer by about six feet. He sailed along, his arms spinning through the air and water like windmills. Michael slowly pulled even farther away from the rest of the swimmers. He was on a **world record** pace!

Michael switches to the backstroke at the 100-meter point.

With a solid lead at 200 meters, Michael begins his breaststroke.

At 200 meters, the swimmers rolled onto their stomachs for the **breaststroke.** Michael glided through the water. With every stroke, his head and shoulders bobbed out of the water.

After 300 meters, Michael was picking up more speed. Swimming **freestyle,** he churned his arms and legs. Right, left, right, left. Michael Phelps was going to win an Olympic **gold medal!**

But could he break the world record? Michael swam as hard as he could until he touched the pool wall. He'd finished the race in 4 minutes and 8.26 seconds—a new world record!

"Yeah!" Michael screamed with joy. His dream of winning an Olympic gold medal had finally come true.

But he wasn't done. He had more events to swim in this Olympics. Michael was trying to do what no athlete had ever done. He was trying to win eight gold medals in a single Olympic Games.

Michael won his first gold medal in the 400-meter medley.

Michael showed talent and determination at an early age.

At Home in the Water

Michael Phelps was born on June 30, 1985, in Baltimore, Maryland. His mother, Debbie, was a teacher. His father, Fred, is a Maryland State Police officer. Michael's parents were divorced when he was in elementary school.

Michael grew up near Baltimore, Maryland.

Michael's two older sisters, Hilary and Whitney, are both talented swimmers. "My older sisters started the family in the sport of swimming," says Michael. "I grew up around the pool. . . . That was normal for me, just hanging around the pool."

Michael joined the North Baltimore Aquatic Club (NBAC) team in 1992, when he was seven years old. He practiced swimming and competed in **meets** against other kids. Michael soon showed that he had a real talent for swimming.

In 1996, when Michael was eleven, he began working with a coach named Bob Bowman. Coach Bowman saw that Michael had a chance to be a great athlete. He encouraged Michael to work hard.

Michael has the perfect swimmer's body. He has long arms and big hands that can really move water. He also has short legs, which keep him from dragging in the water. Michael has big feet too. They work like dolphin flippers to move him through the water.

Michael began to practice twice a day, every day. "I swim seven days a week, two to five hours a day, about 50 miles a week," Michael says. "Once I'm in the water, I feel more at home."

Michael trained with NBAC's best athletes. Most of these swimmers were six or seven years older than Michael.

Before long, Michael began to catch up to the older swimmers. He won race after race. Some of these older swimmers didn't like losing to a young kid. "I got picked on some. . . . I would get frustrated. But it didn't do what they wanted, which I guess was to make me quit. It just made me swim faster."

Michael kept getting better. By 2000, fifteen-year-old Michael was already one of the best swimmers in the country. He was ready to try to achieve every swimmer's goal—to swim at the Olympic Games.

Michael's sister and mother watch him win a race.

In 2000, Michael earned a spot on the U.S. Olympic Swim Team.

Boy Wonder

No one expected Michael to make the U.S. Olympic Swim Team. But he shocked everyone at the 2000 **Olympic Trials.** He swam great and earned a spot on the team. Michael was the youngest male swimmer to make the U.S. team in sixty-eight years!

At the 2000 Olympic Games in Sydney, Australia, Michael swam in the 200-meter butterfly event. He finished fifth, so he didn't earn a medal. He was disappointed but eager to get better.

His time in Sydney gave him a chance to watch some of the world's best swimmers. These swimmers included gold medalists Ian Thorpe of Australia and Pieter van den Hoogenband of the Netherlands. Michael wanted to be a superstar just like them.

Ian Thorpe *(left)* and Pieter van den Hoogenband swim at the 2000 Olympic Games in Sydney.

Michael swims the butterfly at the Sydney International Aquatic Center.

Michael kept training every day. He continued to improve. In early 2001, he set his first world record. He swam the best time ever in the 200-meter butterfly.

But that was just the beginning. Two years later, Michael set a world record in the 200-meter individual medley.

In July 2003, Michael competed in the World Championships in Barcelona, Spain. He broke *five* world records. He even set two world records on the same day! No one had ever done that before.

Just a few weeks later, he broke his own world record in the 200-meter individual medley. He had set seven world records in only a few months!

Suddenly, everyone was calling Michael the best swimmer in the world. But Michael didn't stop working hard. He set his sights on the 2004 Olympic Games. He was ready to make a big splash.

Michael watches the scoreboard at a race at the 2003 World Championships.

The silver (left), gold (center), and bronze (right) medals hang on display in Athens, Greece. Three thousand medals were made for the 2004 Games.

Racing for the Record

Everyone knew Michael had a chance to win several medals at the 2004 Olympic Games. Some even thought he could break the record for the most gold medals at a single Olympics.

In 1972, U.S. swimmer Mark Spitz won seven gold medals. Could Michael top the great Mark Spitz?

Teammates carry Mark Spitz on their shoulders at the 1972 Olympic Games in Munich, West Germany. Spitz won seven gold medals at the 1972 Games.

Michael decided to give it a try. He **qualified** for four individual events. In these he'd swim by himself. He also hoped to swim as part of a team in four **relay** events. "I want to break Mark Spitz's record," said Michael. "But if I can win one, just one [gold medal], I will consider these Olympics a success."

Some people thought eight events would be too much to handle. They said Michael should just focus on a few events. “I don’t think anybody is going to win seven gold medals at an Olympics,” said Ian Thorpe. “I think he’s going to walk away disappointed. . . .”

Australian Ian Thorpe knows the pressures of the Olympic games. He swam in his first Olympics in 2000 at age seventeen. He won three gold and two silver medals.

Michael seems relaxed at a news conference at the 2004 Olympic Games.

By the time the 2004 Olympic Games started, Michael was big news. His name and face were on magazine covers. He appeared on TV commercials and interviews. He was the most famous American at the Olympics. Everyone wanted to see if Michael could break Spitz's record. To do this, he would have to win gold in every event.

After winning the gold medal in the 400-meter individual medley, Michael competed in the 4 × 100-meter freestyle relay. Michael's friend and teammate Ian Crocker was sick and had a bad race. Michael and the rest of the U.S. team did well. But the South African team swam an amazing race.

The South Africans won the gold and set a world record. Michael and his teammates had to settle for third place.

The South African swimmers roar in triumph after their gold medal race in the 4 × 100-meter freestyle relay.

Mark Spitz isn't jealous of Michael trying to break his record. "I'd like to see him do it," said Spitz.

The loss meant Michael wouldn't be able to break Mark Spitz's record of seven gold medals. But he didn't mind. He had already reached his goal of winning one gold medal.

Michael's next race would be the toughest of all. For the 200-meter freestyle, Michael was going to take on Thorpe and van den Hoogenband. They were two of the best swimmers of all time.

Michael (left) and Ian Thorpe stretch before the 200-meter freestyle event. Michael revs up for a race by listening to hip-hop on his headphones.

Great Swimmer, Great Friend

Thorpe and van den Hoogenband were the best 200-meter freestyle swimmers in the world. Some people said Michael should skip the 200-meter freestyle and focus on a different event. But Michael wanted to swim against the best.

Michael, van den Hoogenband, and Thorpe race neck and neck for the gold in the 200-meter freestyle.

In an exciting final, Michael came in third behind van den Hoogenband and Thorpe. He'd lost his chance to match Spitz's record. But he was happy he had tried.

Michael had already earned three medals—one gold and two bronzes. He had already had a great Olympics. And he still had five more events to go.

Next, he won gold in the 200-meter butterfly. Later that night, he helped the U.S. team win an exciting 4 × 200-meter relay. The Americans beat Thorpe and the Australian team by a split second!

Michael was on a roll. Two days later, he won the 200-meter individual medley. Michael had won four gold medals!

Michael and his teammate react when their winning time appears on the scoreboard after the 4 × 200-meter freestyle relay.

Michael checks the scoreboard after swimming the prelims for the 100-meter butterfly.

The next day, he raced against his teammate Ian Crocker in the 100-meter butterfly. The two friends swam a tight race. Crocker had the lead from the start. But Michael was determined to win. He made a great finish to beat Crocker by a fraction of a second. Michael had earned seven medals!

He had one more event to compete in—the 4 × 100-meter medley relay. In swimming, each team can choose any four team members for the relay event. Michael was chosen to swim butterfly in the relay. But he gave up his spot in the final to Crocker. Michael wanted his friend to have another chance to win a gold medal.

"Ian and I are good friends. This is a decision that I chose," said Michael. "I'll be in the stands cheering."

"I'm speechless," said Crocker. "It's a huge gift." Crocker and the relay team went on to win the gold medal. Since Michael had swum in the **prelims,** he was considered part of the team. So he got a gold medal too.

Gold medalist Michael and silver medalist Ian Crocker pose after the 100-meter butterfly. Michael had to swim hard to beat Crocker, who held the world record for the event.

Michael had earned eight medals—six gold and two bronze—in a single Olympic Games! It had been an incredible performance. What would he do next? Coach Bowman said, "I don't even think we're near his limit yet."

Before the Athens games had even ended, Michael was talking about the 2008 Olympic Games in Beijing, China. Would he take another shot at topping Mark Spitz? "A lot of things can happen in four years," said Michael. "I wouldn't count anything out."

"I just want to go home and sleep," Michael said after the 2004 Olympics ended. But within a few weeks, he was back in competition. Michael will continue to compete and train for the 2008 Olympics in Beijing, China.

Selected Career Highlights

2004 Earned six Olympic gold medals, one each in 400-meter individual medley, 200-meter butterfly, 4 × 200-meter freestyle relay, 200-meter individual medley, 100-meter butterfly, and the 4 × 100-meter medley relay
Broke his own world record in the 400-meter individual medley
Won Olympic bronze medals in 4 × 100-meter freestyle relay and 200-meter individual freestyle
Set Olympic records in the 200-meter butterfly and 200-meter individual medley
Set a world record in the 400-meter individual medley during U.S. Olympic Team Trials in Long Beach, California

2003 Broke five world records at the 2003 World Championships in Barcelona, Spain
Set a world record in the 200-meter butterfly at the World Championships
Finished first in the 400-meter individual medley at the World Championships
Set world records twice in winning the 200-meter individual medley at the 2003 World Championships
Finished second in the 100-meter butterfly

2001 Became youngest male swimmer ever to break a world record

2000 Was the youngest male swimmer to make a U.S. Olympic team since 1932
Finished fifth in the 200-meter butterfly at the 2000 Olympic Games in Sydney, Australia

Glossary

backstroke: a stroke in which swimmers stay on their backs and pull their arms through the water one at a time. The legs kick separately up and down, called a flutter kick.

breaststroke: a stroke in which swimmers stay on their stomachs and pull their arms together beneath the surface of the water. Their legs kick together.

butterfly: a stroke in which the swimmer swims on his or her chest. The arms move together over and through the water. The legs kick together in a dolphin kick.

event: a race where swimmers compete in a stroke over a certain distance. For example, the 200-meter butterfly is an event in a swimming meet. An event can include prelims and finals.

favorite: the person most likely to win

final: the race that determines the winner of an event. Swimmers reach the final by doing well in the prelims and semifinals.

freestyle: a race in which swimmers can use any stroke they want. Freestyle swimmers today use the front crawl, the fastest stroke. In the front crawl, swimmers glide face down and pull their arms through the water one at a time. The legs kick separately, up and down.

gold medal: in the Olympics, the award for first place in an event. Second place receives a silver medal. Third place wins a bronze medal.

medley: an event in which a swimmer or a relay team swims a variety of strokes—the butterfly, the backstroke, the breaststroke, and freestyle

meets: gatherings where swimmers race each other

Olympic trials: a meet held a few months before each Olympic Games to determine who will make a country's Olympic team

prelims: short for preliminaries, the early races in a swim event. The best swimmers in the prelims go on to swim in the finals.

qualified: to have earned the right to swim in an event

relay: an event in which team members take turns swimming, one after the other. For example, in a 4 × 100-meter relay, each of four team members swim 100 meters.

starting block: a platform at the edge of the pool that a swimmer dives from at the start of a race

stroke: a certain style of swimming. For instance, the butterfly and the backstroke are swimming strokes.

world record: the fastest time ever recorded in an event

Further Reading & Websites

Edelson, Paula. *Superstars of Men's Swimming and Diving.* Philadelphia: Chelsea House Publishers, 1999.

Oxlade, Chris, and David Ballheimer. *Olympics.* New York: DK Publishing, Inc. 1999.

Rouse, Jeff. *The Young Swimmer.* New York: DK Publishing, Inc. 1997.

The Official Website of the Olympic Movement
http://www.olympic.org/uk/sports/index_uk.asp
Learn more about the sport of swimming in this website's "Aquatics" section.

Official Site of Michael Phelps
http://michaelphelps.com
Michael's website has the latest news, pictures, and a short biography of Michael. You can even e-mail a question for him to answer.

Sports Illustrated for Kids
http://www.sikids.com
The *Sports Illustrated for Kids* website covers all sports, including swimming and other Olympic events.

Index

Photo Acknowledgments

Photographs are used with the permission of: © David Gray/Reuters/Corbis, pp. 4, 21, 29; © Erich Schegal/NewSport/Corbis, p. 6; © Stefan Matzke/NewSport/Corbis, p. 7; © Icon SMI, pp. 8, 12, 16, 19, 24, 28; © Duomo/CORBIS, pp. 9, 13; © Royalty-Free/CORBIS, p. 10; © SEGUIN FRANK/CORBIS SYGMA, p. 14; © Getty Images, p. 15; © YIORGOS KARAHALIS/Reuters/Corbis, p. 17; © Bettmann/CORBIS, p. 18; © Jeff Christensen/Reuters/Corbis, p. 20; © MIKE BLAKE/Reuters/Corbis, p. 23; JERRY LAMPEN/Reuters/Corbis, p. 25; © CHARLES PLATIAU/Reuters/Corbis, p. 26; © YVES HERMAN/Reuters/Corbis, p. 27.

Cover: © 2004, The Baltimore Sun. Used by permission.